DYLAN'S™ BOOK OF NOISES

LEARN WITH THE MAGIC ROUNDABOUT™

OXFORD
UNIVERSITY PRESS

Zebedee goes . . .

BOING!

BOING!

BOING!

but Dylan just sleeps
the whole night through
and most of the morning too.

Dougal goes . . .
BOW
WOW

but Dylan just sleeps
the whole night through
and most of the morning too.

Florence goes . . .
YOO HOO
YOO
HOO

but Dylan just sleeps
the whole night through
and most of the morning too.

MOO MOO
Ermintrude goes . . .
MOO MOO

but Dylan just sleeps
the whole night through
and most of the morning too.

STOMP!
STOMP!
Soldier Sam goes . . .
STOMP!
STOMP!

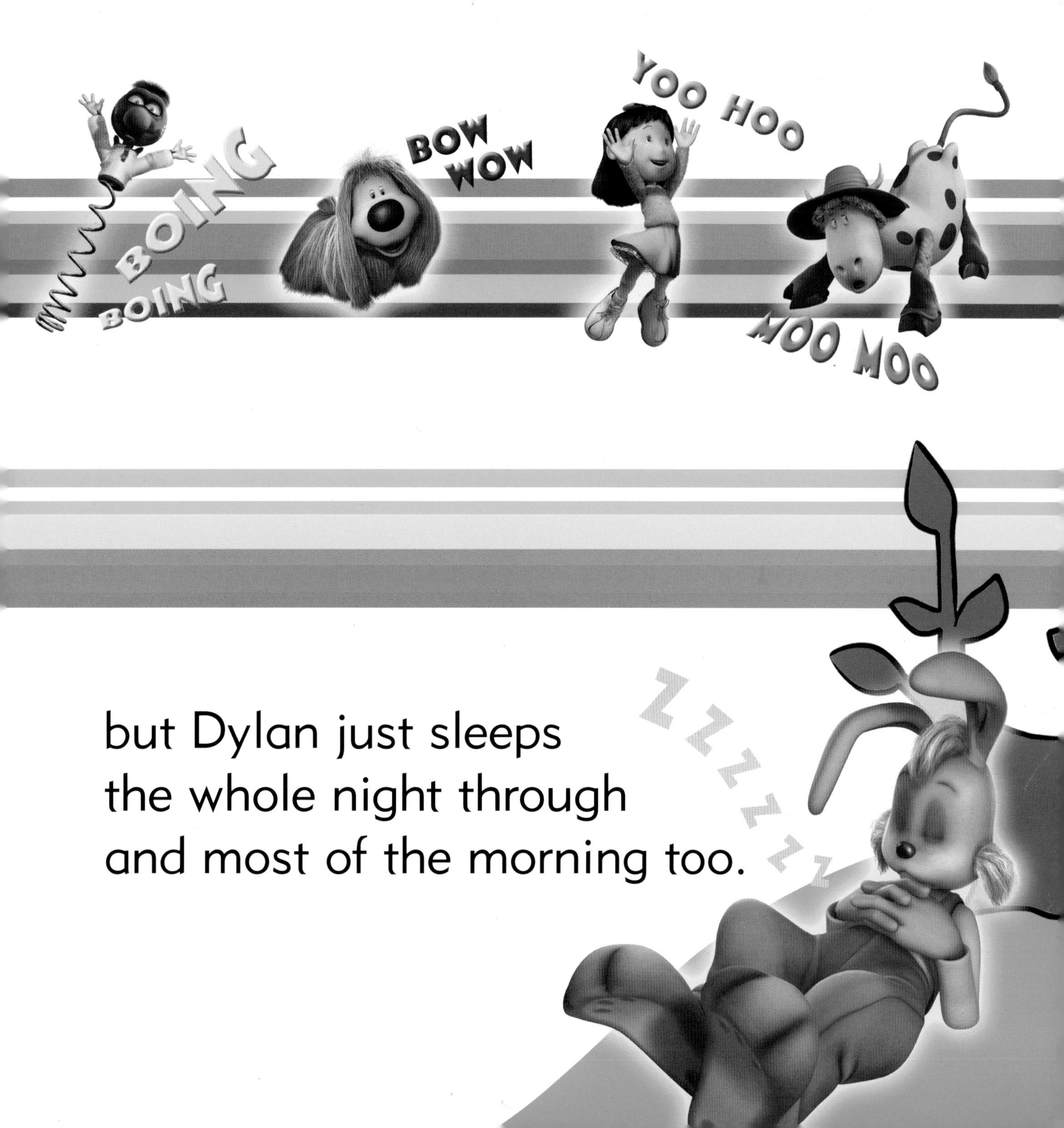

but Dylan just sleeps
the whole night through
and most of the morning too.

TOOT! TOOT!
Little red train goes . . .
TOOT!
TOOT!

but Dylan just sleeps
the whole night through
and most of the morning too.

Zeebad goes . . .
HA!
HA!
HA!
HA!
HA!
HA!

but Dylan just sleeps
the whole night through
and most of the morning too.

SQUIDGITY SQUIDGE

Brian goes . . .

SQUIDGITY SQUIDGE

but Dylan just sleeps
the whole night through
and most of the morning too.

YOOHOOOOOOOOOO
BOW WOW
TOOT TOOT
MOO MOO
SQUIDGITY SQUIDGE

BOING!
BOING!
BOING!
STOMP!
STOMP!
HA!
HA!
HA!
Z
Z
Z
Z
Z

BOING!
BOING!
BOING!
BOW
WOW
YOO HO
HEY!
'Hey, what's that noise?
Hey, what's the score?'
HEY!

'It's time for bed,' says Zebedee.
'This noise is just too loud for me!'

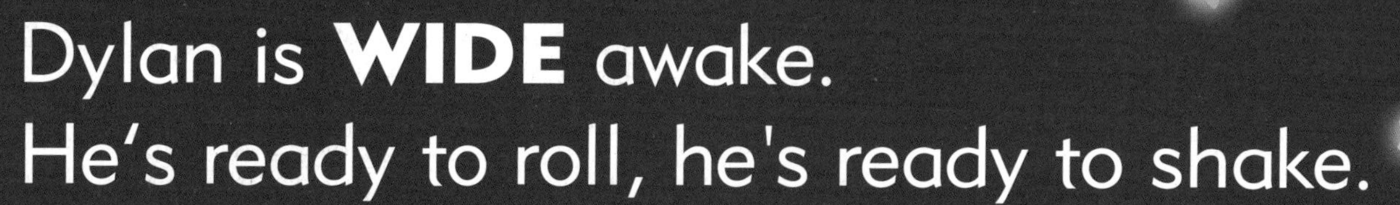

Dylan is **WIDE** awake.
He's ready to roll, he's ready to shake.

But his friends want to sleep
the whole night through,
and wake up in the morning
just like you!

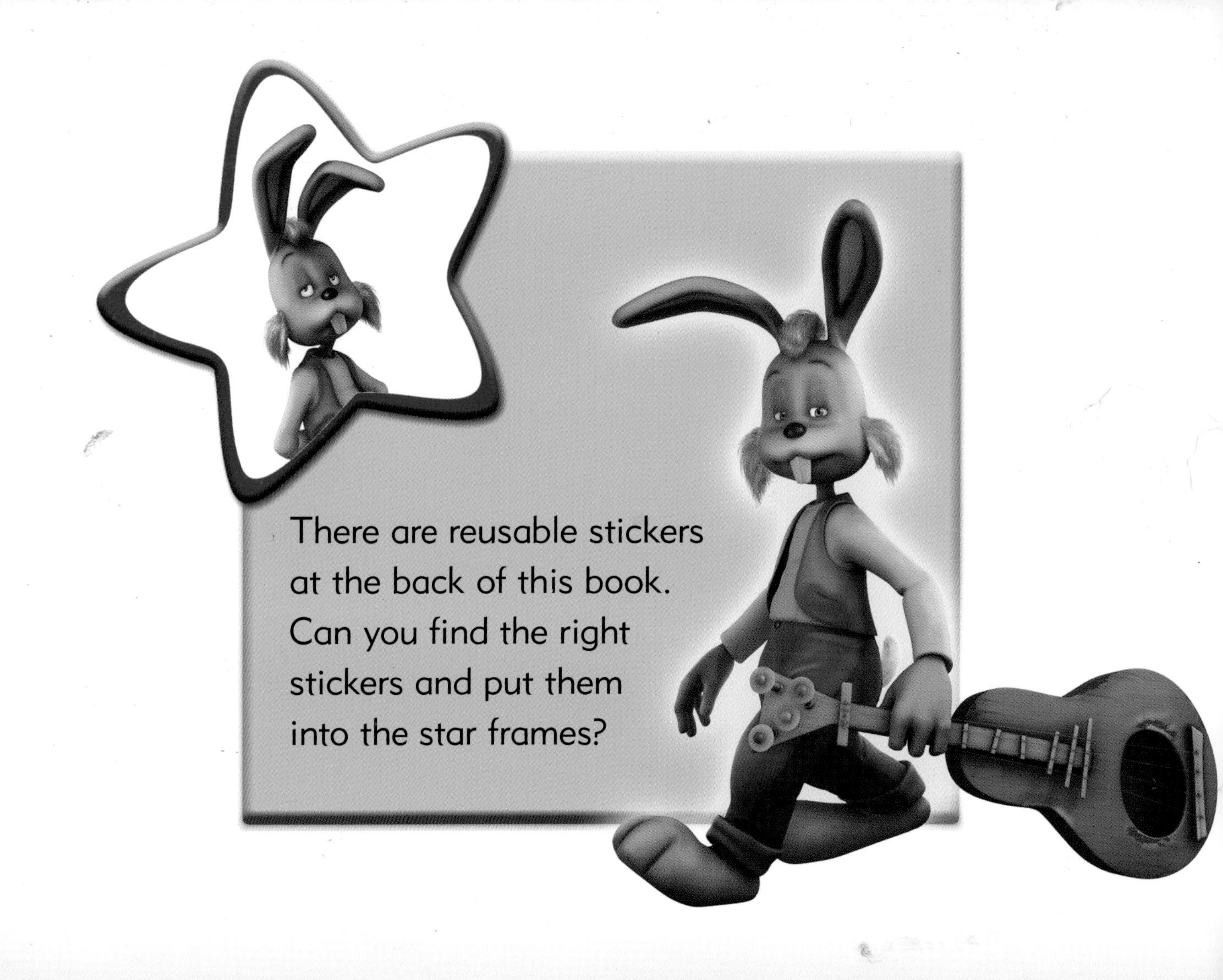

OXFORD
UNIVERSITY PRESS
Great Clarendon Street, Oxford OX2 6DP Oxford is a registered trade mark of Oxford University Press in the UK and in certain other countries

Database right Oxford University Press (maker) First published 2005 by Oxford University Press All rights reserved. British Library Cataloguing in Publication Data Data available ISBN-13: 978-0-19-911276-0 ISBN-10: 0-19-911276-2 1 3 5 7 9 10 8 6 4 2 Printed in China

www.the-magic-roundabout.com